SCOUTING FOR EXTRA-TERRESTRIAL LIFE

BIPIN MENON

ISBN 979-888606844-3

Contents

Preface

This book seeks to arouse scientific curiosity on Man's quest for scouting life beyond our planet. It highlights the developments related to sighting of UFOs and alleged encounter with aliens. The statements by the key political leaders and scientists on this topic provides a background for readers to understand on where we stand in terms of finding extra-terrestrial life. The bizzare cases of alien kidnappings does not seek to sensationalise the topic but is aimed at putting into perspective some of the unexplained events. Man's progress on space exploration has been delved into in terms of the spaceships sent beyond our solar system, the objects such as Oumuamua which has traversed the sun and our planets and the alignment of the James Webb telescope which would open up the hitherto unexplored space. There are chapters dedicated to the possible bodies in the solar system that may harbour microbial life. The inferences and conclusions seek to provide a synopsis of the current scientific developments and how we should be ready to explore the vast universein search of our quest for extra-terrestrial life. There are exciting times ahead as we seek to go and observe space as never before.

CHAPTER ONE

Introduction

Man's quest for discovering extra-terrestrial (ET) life forms, i.e. any life beyond our planet has been etched in human history. A lot of literature has used the term "*alien*" to describe these entities and technologies that have a source beyond our planet. The topic has generated a lot of interest not only among the professional astronomers who peek into the sky with their state of the art equipment but also a lot of amateurs who have been fascinated with what lies beyond our planet.

Carl Sagan, the famous astrophysicist had mused quite philosophically that

"In the deepest sense the search for extra-terrestrial intelligence is a search for ourselves."

All this interest has stemmed from the fact that with countless stars and their planetary systems how is it possible that there has not been an iota of credible and verifiable evidence of extra-terrestrial life.

Warner von Braun, the aerospace engineer had remarked

"Our sun is one of 100 billion stars in our galaxy. Our galaxy is one of the billions of galaxies populating the universe. It would be the height of presumption to think that we are the only living things within that enormous immensity."

There would surely be bodies that would mimic the conditions of our planet and hence ensure that life, as we know it, thrives. Many call it the habitable zone of our universe where there is possibility of sufficient atmospheric pressure, availability of water and presence of organic molecules that may sustain life. By the theory of probability, even the die-hard skeptics are surprised that not one concrete shred of evidence exists in public domain of such life, in whatever form it may be. One could argue that the 2019 US military admission of certain unknown spacecraft tracking their warships could be the first such instance, but there has not been any other information available publicly. This has also lent credibility in the minds of some to the various sightings made of unidentified flying objects (UFOs).

The Italian scientist Enrico Fermi has postulated his paradox theory in 1950 when he stated that any civilization with time would be able to have its footprint across the universe. He said that our solar system is relatively a newcomer with an age of around 5 billion years as compared to the universe which is around 14 billion years. Hence, it is possible that either ET's never came or if at all they did, it would have happened much before civilization started.

The entertainment industry has given a completely new twist to this tale by portraying ET beings in a form slightly different from human beings, sometimes known by the term humanoids. There are of course variations of this form including those with the face akin to reptiles replete with claws and pincers. Many speculate that the reason was that the film industry wanted to economise costs for the protagonists to don their costumes. Any shape other than humanoids would have entailed a higher cost element and weighed down on the already tight budgets of the entertainment industry. Of course, this hasn't stopped the creative team from experimenting with some appendages of reptiles, spiders, octopuses, bugs etc.

Where did these forms emanate from? There is of course the scientific explanation on how could intelligent life exist in other forms, since they would need earth like conditions and would evolve like we have done on this planet. Yet another consideration is that possibly the structure of a humanoid is more amenable, atleast in our current thought process, to effectively use tools, be adequately mobile in the bipedal mode and have a proportionately developed brain. Another school of thought is that other shapes would have confused the audiences and lowered the relatability quotient. Some argue that these shapes may also have emanated from the actual sightings reported, but never verified. However, this could be the classic "*chicken and egg*" situation, did the films lend credence to the humanoid form which led to those sightings or vice versa?

Even the vehicles supposedly used by them are mostly in a circular disc form known as saucers which is quite different from those on earth. The genesis of this can be traced to a newspaper article in 1947 of reported sightings of saucer shaped objects by a pilot Kenneth Arnold in Washington State. Arnold gave a series of interviews wherein his statements on the nine lights that he saw as saucers floating over water, being interpreted as saucer shaped vehicles. Some still suspect that these objects may have been technology well ahead of its day, either from the US military designers or even some other species. Strangely even the top notch automobile engineering and space exploration enterprises are yet to design a commercially viable vehicle of this shape. The conventional laws of physics don't render these shapes aerodynamic and hence have not really seeped into automobile design. However, for inter-stellar space travel it is shape agnostic, i.e. spacecraft design does not matter in the absence of air drag. There are two explanations for these sightings, if they are indeed true. The secret military aircraft may be the cause of these sightings though they still defy our current understanding. However, some of the modern machines are coming closer to these shapes. A more far-fetched theory would of course be that they are advanced technologies that probably even defy gravity and hence the shape does not matter. A saucer shape may possibly be more convenient providing a 180 degree panoramic view for its occupants.

There have been a number of reported sightings of both UFOs and aliens. Even bizzare cases of abductions by aliens and transportation of people to other planets and systems have been reported. While none has been really verified, sometimes it is difficult to fathom certain cases where the person concerned spoke about phenomena on which no publicly available information was available or they passed lie detector tests. Many investigators including law enforcement officers and medical professionals were of the view that the persons believed strongly in what they reported.

Moreover, the various purported sightings that have been made across the globe seem to have embraced Hollywood's view of these ETs including their saucer shaped vehicles. With the advances in digital technology of photoshopping, many of these images and videos seem real and could spook anyone. There is no doubt that the subject is of such a high magnitude of interest globally that there is a big school of believers who would go to any extent to justify the existence of ETs.

However, to the credit of the entertainment industry, it has been their effort which has really generated public interest in this topic. The industry may have made millions with their films but they have instilled this sense of curiosity and awe in even in the most scientifically disinterested denizens of this planet.

On top of that there has been a lot of speculation even emanating out of some renowned military and political figures which has aroused a lot of suspicion. While the skeptics say that these statements emanate out of the need for publicity, others speculate that the top bosses are hiding something from the world of this ET phenomena. A weirder theory is that ETs may actually be living among us even today without us noticing since they are playing the perfect masquerade, given their higher intelligence forms. Some even speculate that these advanced ETs may actually controlling us and could thus easily erase our memories of these events. A school of thought that we may actually be objects of an experiment thus creates a spaghetti bowl of possibilities

Whatever, be the real story, the topic has aroused a lot of interest. Science has been making its best efforts to make contact with any other forms of life, both advanced and primitive. There have been some cautionary words fro some like Stephen Hawking that there could be a host of predatory aliens who might then try to conquer us. Some strange radio signals have been received from inter-stellar space but no one has been able to decipher it. Some of the satellites like Voyagers 1 and 2, Pioneers 10 and 11 and New Horizon have crossed our solar system but have not communicated any credible piece of information which could arouse suspicion. Nevertheless, they have come across some objects like the rotating Arawn, which are yet to be analysed. The fastest of these spacecraft, the Voyager 1 is the farthest from earth but is likely to reach the next star in another 40,000 years. Moreover, communication has been lost with some of these probes and some of them are running out of energy sources thus making it difficult, even if any ET is able to contact them.

It is also important to note that in human history, we have made serious efforts at space exploration only in the last 60 years. A major barrier that exists today is the threshold that we are not able to even reach a fraction of the speed of light in inter-stellar travel and hence have to rely on the faster radio signal communications. There is a lot of debate on creation of a space economy as we enter this next phase. A global collaborative effort may be needed for this as we expand beyond the only habitable planet we know, including in search of ET life and civilisations.

This book seeks to explore this subject in greater detail including looking at the incidents purported to have come from other life forms outside our planet. Some chapters have been dedicated to the planets and moons in our solar system where the possibility of life exists, maybe even below the surface in the oceans and other large water bodies that are predicted to exist there. Rather than sensationalizing ETs and aliens, the aim is to instill scientific interest and temperament as we go into the next era of scouting for life beyond our planet.

CHAPTER TWO

Statements on UFOs and alien life

A number of prominent personalities have spoken about UFOs and alien life. However, there are claims and counter-claims that these may not be authentic and more to do with personal publicity or some entrenched views on the subject not supported by concrete evidence.

Stephen Hawking, a famous physicist, also opined on the ETs and how the interaction with man could actually be counterproductive. His views, as expressed in a documentary series "*Into the Universe with Stephen Hawking*" were that the if ETs were advanced enough to contact us, they could also be dangerous. This could be manifested in them seeking to conquer and rule over the planetary systems. He was, like many others, enamoured with ET life and stated that it would be the greatest discovery in history. He also evinced a lot of interest in the "*Breakthrough Starshot*" project which sought to use propulsion technologies like light sail which would enable crafts to touch a fraction of the speed of light for reaching the nearest Alpha Centauri star system within a human lifespan.

In the context of US Presidents, Gerald Ford said that when he sought information on UFOs, the officials strangely denied the UFO allegations. Bill Clinton and Barrack Obama had the same story indicating that there was complete official denial to the existence of any aliens or their spacecraft which may have been captured.

Even the former US President, Jimmy Carter along with others in 1969 at Leary, Georgia, saw a bright light which changed colours and drifted off. However, the investigations pointed to possibility of it being the halo of Venus or some military aircraft. When questioned on it, President Carter said that while he did not know what it was, he would definitely take any UFO sighting incident seriously.

The Canadian defence minister during the 1960s; Paul Hellyer has been a strong believer in aliens and has talked extensively on the subject. In his interviews, chiefly one in 2014 given to the Russian channel RT, he had stated that there are around eighty species of aliens who were also among us categorizing some of them being five feet tall while others being the tall whites. While stating that around 80% of the UFO sightings may be fakes, he was of the view that there were some genuine reporting's of craft with technology much beyond our comprehension. Hellyer further stated that advanced alien technology such as microchips and kevlar were earlier shared with humanity. His view was that they were benign and would share more if we took care of the environmental deprivation of our planet and not use atomic weapons. Coming from a renowned politician who was also a pilot and engineer, this is indeed something to take note of. However, whether it is credible is still an open question, since Hellyer himself had never met the ETs, purported to be on earth. Moreover, he claims to have seen one UFO. Isn't it curious that if he knew so much, he would have made efforts to find out and take pictures of atleast some of the species of ETs?

Henry Kissinger, the US Secretary of State had mentioned that "*the subject is the biggest hot potato of all time*".

Edwin Aldrin, the second man to set foot on the moon in the Apollo 10 mission of 1969, reported seeing a UFO. He even passed a lie detector test on his sighting.

Haim Eshed, former head of the Israeli space agency had on 8 December, 2020 said that the US was in touch with alien civilisations through a "*Galactic Federation*". A respected figure in his profession, these startling revelations were shocking. He even stated that the former US President, Donald Trump was on the verge of spilling the beans to the public before being asked not to do so for fear of creating a hysteria. It was mentioned that the world was not yet ready for such a revelation. He stated that there is an agreement of the Federation with the US government wherein there are underground bases under Mars where both sides meet. While these statements attracted a lot of interest, it was also a subject to a lot of ridicule. Some questioned the timing as it came on the back of a book based on the interview given by Eshed titled "*The Universe Beyond the Horizon*". Many said that this was just a publicity stunt for the book. This may well seem to be one of the unverifiable stories like that of the UFO. However, what is surprising is that Eshed is an accomplished expert in space exploration and the statements coming out of him would always leave that ray of doubt on whether it was indeed concocted.

Bill Nelson, the chief of NASA during an October, 2021 interview stated that the aim of his organization was to seek out life beyond earth. While dwelling on some of the UFO encounters by the military agencies, he surmised that some of these objects seemed to have maneuverability which was extraordinary. He also mentioned that in his interviews with some the pilots who saw UFOs, he was convinced that they did see something. He ended his note with an unusual admission of possible life beyond our planet with the statement that "*Who I am to say that planet earth is the only location of a life form that is civilized and organised like ours?*"

Former US President Obama in an interview in May 2021 stated that there is little explanation to some of the encounters. He also mentioned that he could not speak on aliens publicly. Some may interpret this as an admission that aliens exist. However, it could also be a ploy to keep interest alive in the subject and more importantly keep your potential adversaries uncertain.

When former US President Bill Clinton was asked in an interview in June, 2021 about the mysterious encounters, he stated that given the vastness of the universe, there was a high probability of life beyond earth. He also mentioned that there were a number of UFO encounters which haven't been fully understood. Didn't look too different from what the NASA chief said on the subject? Could it be the same source of debriefings?

The former US President Donald Trump had jumped into the UFO fray and had a lot of interest in the subject even before he took office at the White House. However, subsequently when he was interviewed as President, his body language indicated a lot of disinterest on the subject and the responses to the questions raised were lukewarm, a far cry from his flamboyant style. These may be possible cues that there may have been no significant discovery of ETs or UFOs, worth all the brouhaha being generated globally.

Finally, what do we make out of all these statements coming from prominent personalities? Very difficult to assess given that many would have been adept at masking their body language and inner feelings while speaking on the subject. Comments from famous scientists like Eshed may have to be taken with a pinch of salt given the background of his book publicity. As for the US presidents, it would have been difficult to keep any major discovery such as contact with an advanced civilization, a state secret given the transparency in that country and the fact that any such aliens would then be capable of controlling us, rather than the other way round. The former Canadian defence Minister's statements also have to be seen in the light of the fact that he himself saw only a single UFO without having a sight of any of the 80 species of aliens that he claims live on earth.

However, this does not rule out discovery of microbial or primitive life forms somewhere, but then it does not make sense to keep these a secret from the world.

One question that arises is to why all this is not being put out in the open. Is it basically a ploy to keep your adversaries guessing in the game of one-upmanship on new military technologies being developed? Therefore, the mystery of whether we have or have not found any alien technology, remains a strong psychological weapon in the new cold war era.

CHAPTER THREE

Military and law enforcement sightings of UFOs

Before, we dwell on the general sightings of UFOs, let us focus on what we believe are more credible ones. One category of sightings by military and law enforcement personnel, be in an aircraft, ship or on land becomes more difficult to rebut. This is on account of them having the most modern transport vehicles and their own physical fitness being of the highest levels. However, since they involve the military, one cannot rule out the possibility of secrecy being maintained by the powers that be so as not to create any panic. But despite this hypothesis, one cannot rule out some other factors influencing the sightings.

On 1 October, 1948, a dogfight in a P-51 Mustang occurred with a bright object in the night sky over Fargo, North Dakota, US. The pilot George Gorman chased the object which is supposed to have outsped him and undertook maneuvers beyond that of his plane. The light was also spotted by another plane pilot along with a passenger and an air traffic controller over Fargo airport. The investigative report however, put it as a weather balloon and possibly the planet Jupiter but which still did not explain the dogfight maneuvers that Gorman, an experienced pilot made. Hence, it still remains a mystery.

In the first of its case where the air force of a country was involved, occurred during April, 1969, when the Finnish air force was alerted on seven balloons. The aircraft which went to intercept then reported these being round or disc shaped. However, when approached, the objects accelerated away. This was later attributed to some spy aircraft of other countries. The credibility of this theory was high given the cold war scenario.

Probably, the first major UFO incident in the Middle East occurred in September, 1976 in the skies over Teheran. When two F-4 Phantom fighters were sent to intercept the object which gave off a bright light, they were purported to have lost their communication systems when they approached it. Only after withdrawal from the zone of combat, did these systems revive. One of the fighter pilots also attempted to fire a missile after locking the target but his system shut down. On the contrary, the second pilot claimed that something was fired at him by the object which crippled his communications. Moreover, the second pilot saw a light crash to the ground but when he subsequently investigated the next day, but could not find any trace where it was purported to land. A series of investigations came to the conclusion that only one of the airplanes had an electrical failure, which was not infrequent given its flying history. Moreover, the pilots could have mistaken Jupiter to be the light source. Another theory was that it was the time of meteorite showers, the Gamma Piscid and the Southern Piscids and the lights must have been these showers.

A Minnesota Deputy Sherif Val Johnson was on a night patrol in August 1979 when he encountered a bright light that moved towards his car, engulfed it and knocked him out.

When he woke up, the car was damaged and his watch and the car clock had also stopped. The car was found sideways on the road with its windscreen shattered and damage to its aerial and headlights. Being a law enforcement officer, many believed that the incident needed further probe while others thought that it was a hoax.

In November, 1980 at the Arequipa region of Peru, a Sukhoi aircraft was sent to intercept a UFO and the pilot fired a large number of shells at it. The pilot reported it was creamish colour with features of a light bulb. However, he had to give up the chase midway.

On 5 March, 2004, Mexican fighter pilots searching for drug smugglers aircraft in the Campeche area near the Gulf of Mexico saw eleven lights. These were later suspected to be flares from offshore oil platforms though the Mexican investigation still puts them as UFOs.

In 2005, two Slovakian jets flying over the Bohunice Nuclear Power Plant encountered fast moving objects. However, this was not picked up by ground controllers. Later on in 2014, a Slovakian cargo pilot over the Zilina region picked up a rocket. However, when checking if there were any military bases from which the object could have emanated, there was no confirmation

It is interesting to see the set of public videos on sighting by the US warships in the 21st century. On 14 November, 2004, fighter planes aboard the USS Nimitz in South California recorded an aircraft in their radar. However, they had been preparing for a fortnight on account of sightings of unusual air movement in that area. A decade later in 2014, aircraft aboard US Theodore Roosevelt on the east coast also recorded in their radar some UFO going at fast speeds. A triangular shape was noted by the USS Russel in 2019 and this picture was released in April 2021. A lot of speculation also went on the possibility of Chinese or Russian spying through drones with a view to looking at the US's capability to respond. There was a sighting in July, 2019 by the US navy in San Diego of a mysterious aircraft over the sea that then plunged into it. It was following the warship USS Omaha. The radar of the incident did surprise even the hardened military personnel, though one must admit that in the video, the commentators tone looked more like a sarcastic laugh than a state of shock.. These events involving the most power military power of the world are arguably the most potent arguments that these sightings need to be taken seriously. However, even here, some skeptics may argue that all these could be some military testing which may have been kept a secret or even some terrestrial adversary trying to spy. But it still begs the question as to why would the US navy release it? Couldn't it be also a sign of showing your weaknesses in not being able to counter it? However, one could be argued that military strategies could be very complex for layman like us to really understand.

The Unidentified Aerial Phenomena (UAP) report of June, 2021 by the US government which analysed all this data came to the conclusion that the sightings could be attributed to many causes. Some of these were airborne clutter, atmospheric phenomena, US development technology, foreign craft or the "others" category. The last is of course part of the extra-terrestrial (ET) folklore.

We then come to the UFO sightings by astronauts and cosmonauts. Some of the purported sightings such as the Apollo 11 mission to the moon or those by Gemini 7 and Gemini 4 were either a hoax or some unidentified space debris. Even the inhabitants of the International Space

Station (ISS) have seen some UFOs but these could never be conclusively attributed to something extra-terrestrial. Ivan Wagner, a Russian cosmonaut while on the ISS in August, 2020 filmed a one minute video of the Aurora Borealis over Australia wherein some unidentified lights were also seen. His tweet interestingly provided three options, meteors, satellites or? Of course this was adequate fodder for a number of people to jump in with their theories.

While all these military and space sightings cannot be just wishy washed away as human errors due to some phenomena, one must keep in mind that there is no publicly available solid proof that they were indeed ET. Moreover, even if there is some clue available, it is unlikely that the military would make it public without political approval. There is of course the logic that such reporting's also make perfect sense to keep the other military powers uncertain on your capabilities, especially if you are building advanced technologies.

CHAPTER FOUR

Bizarre Reporting of Alien Abductions

The story of UFO gets bizarre with many of alleged tales of abductions. While this makes for perfect thrillers, many people remain skeptical of the veracity of these incidents. One school of thought, though in minority, professes that these are plausible incidents and given their advanced life forms, there would be little traces left for any investigation. On the other hand, most investigators are of the view that these are hoaxes perpetrated for publicity such as selling books or earning endorsement reviews. Some point out that it may be the result of dreams or hypnosis. Whatever, one may say, the narration of these alleged incidents makes interesting fodder for the propagation of the UFO suspense and keeps the public curiosity alive.

Interestingly, most of these sightings point to the aliens having a humanoid form and the vehicle being in the shape of a saucer, quite reminiscent of how Hollywood would like us to believe. This definitely arouses more suspicion as to why these forms cannot be like us or other living creatures on earth. The investigations made of these incidents have also not thrown much light on these. We would now discuss some of these bizarre incidents reported in history which have captivated the interest of the common man.

In 1954, two men, Gustavo Gonzalez and Jose Ponce in Venezuela reported that they were driving when a sphere came from the sky and blocked their path. Three humanoid aliens, who were hairy then tried to abduct them. The incredible part of the story is that Gonzalez fought these aliens with a knife but could not hurt them. He then mentioned being paralysed by a beam of light and the craft took off.

Arguably the most bizzare of the UFO stories emanates in July 1956 by Elizabeth Klarer, a South African resident who claimed that she had gone to the planet Meton, which was orbiting Proxima Centauri, and was inhabited by the Venusian civilization. Interestingly, she had been attached with the Royal Air Force and the British military intelligence and was an expert in aviation.. The encounters with the spaceship occurred in a hill known as the Flying Saucer Hill. During the second encounter with the spaceship in the Natal midlands, she was taken into the spacecraft and transported to the mother ship. After reaching Meton which took 4 months, she is supposed to have married one of the aliens Akon and even had a child named Aylin. She wrote a book called "*Beyond the Light Barrier*" where she explains these encounters in detail. It seems on account of health issues, she had to return from Meton. While some experts believed some parts of her story, most debunked it stating that it was a wild imagination without any substance. However, it was the first reported instance of an alleged contact with an alien civilization.

A strange case of an external implant by aliens was that of Jesse Long in Tennessee, US. While claiming to have his first abduction in 1957, he mentioned that an implant was put in his left leg. This implant was removed in 1991 and found to be of a strange surface composition. He also stated that during his later abductions, he saw his nine hybrid children who were the result of the sperm drawn from him by the aliens. He stated that it was a painful procedure and he still wonders why he was part of this breeding programme. Some medical specialists also believe that he had gone through some trauma with some stating that it might be a disorder.

Another story, this time of an abduction by aliens was reported by a Brazilian farmer Antonio Boas in October 1957. The story is about how a strange craft with four humanoids landed on his farm and abducted him. The interesting part of the story is that they then made him have sexual intercourse with a female humanoid. After that the story goes that he was given a tour of their ship and then left back on the ground, a full four hours after he was allegedly abducted. He believed that this was to ensure the creation of a new species. The story had its set of takers but most however remained skeptical of the incident. While Boas developed some medical issues such as nausea, headache, burning sensation and lesions; a medical examination revealed that he was exposed to excessive radiation and these were symptoms of the same.

The Hill incident involving a couple Betty and Barney Hill in New Hampshire in September, 1961 was an alleged abduction by, what they stated were humanoids. However, they had no recollection of this abduction which occurred on the highway as they were driving down from a vacation. On spotting these humanoids, they are supposed to have gone into trance coming back around 56 km from where the alleged abduction took place. It was only through a series of dreams that Betty had subsequently and hypnotic sessions undertaken by the couple that the story of a possible abduction and medical examination by the humanoids of the couple was brought to light. However, with little evidence, there is a lot of skepticism on this story too. There was also a time gap between the occurrence of the incident and the arrival of the couple at home for which no plausible explanation is available.

Another bizzare incident of an alleged kidnapping occurred in October, 1973 when two persons Charles Hickson and Calvin Parker were fishing on the banks of the Pascagoula river in Mississippi, USA. They reported that an oval object then appeared with three robotic creatures having tubular ears, eyes and pincers taking them to their craft for examination. While Parker was silent on the event, the entire incident has been narrated by Hickson. Both of them were subjected to a lie detector test but all three results did not show that they were lying. Even a medical examiner Dr. Julius Bosco was of the view that both believed what they saw and were honest to report it. Another witness, Larry Booth also said that he saw a round object with lights and a dome moving away. Parker came out later with his story that he too was examined by a fourth feminine looking creature inside the craft. The incident still remains a mystery and many who examined the two were of the view that they did experience something that sent them into a state of shock.

Later in November, 1975, a forestry worker, Trevor Walton is alleged to have been abducted by aliens near Snowflake, Arizona. Along with six other workers, they were traversing the Apache-Sitgreaves National Forest when they encountered a gold coloured saucer shape object which was hovering. When Walton went out of curiosity to examine it, he was struck by a white light and became unconscious, as his co-workers ran off in fright. They reported the incident to the sheriff and a large scale search was carried out which could not trace Walton. Walton recounted opening his eyes in an examination room where there were three short bald creatures of humanoid shape. Despite his struggle, his story goes that he was knocked out with a plastic mask. His recollection was that he was left on the roadside with the saucer flying off. What was surprising was that Walton came back after 5 days during which a search was undertaken for him with investigators suspecting his co-workers of murdering him. This was on account of an earlier incident involving Walton and another co-worker. Even the polygraph test carried on him and some of his co-workers did not point to any suspicion that it was a hoax. However, they were the usual doubters who believed that they faked the polygraph test and concocted a story about the abduction. A movie was made of this incident and the seven people financially leveraged the incident. However, many of the investigators were of the view that the men truly believed what they saw and were in a state of shock.

Another alien abduction was reported in Emilcin, Poland reported in May 1978. This time it was the case of a farmer Jan Wolski who was driving his horse cart down a road. He reported that two green faced humanoids jumped out of nowhere and sat next to Wolski asking to be taken to a clearing where a large craft was hovering. He described it as a white object, the size of a bus and hovering around five metres above the ground. When he went inside, he was asked to undress and then medically examined by the humanoids using a saucer shaped object. He reported that the aircraft was dark in colour with no lights or windows. After this incident, there is another reporting, this time of a 6 year old boy of a craft of similar description in that area which lifted off into the sky. Strangely, there was little investigation of this incident and even a newspaper interview given by Wolski was de-classified much later.

An attempted abduction was reported in November, 1979 from Dechmont Law, West Lothian, Scotland when a forestry worker Robert Taylor wandered into the woods. He claimed to have seen a large dome in a clearing hovering above the ground. He then narrates that smaller objects then seized him and tried to take him towards the larger object and in the struggle, he lost consciousness. He is said to come back home in disheveled clothed which were also muddied. In a subsequent investigation of the area, some marks were also found. There were explanations given later of possible PVC pipes laid on the ground by the local water authority that gave the markings. Some attributed Taylor's narration to an epileptic attack wherein he was hallucinating while others stated that it could have been a mirage of Venus or a water tower that led to this story.

In December, 1985, an accomplished writer Whitley Streiber claimed he was abducted from a remote location and penned it down in his novel called Communion. He claimed that the aliens who abducted him had triangular heads, large eyes and were grey in colour. Many claim that this was just a publicity stunt for selling his book.

An incident occurred in December, 1987 at the Ilkey moor in West Yorkshire England, when a retired police officer is said to have seen a short green coloured alien who he photographed. While running after the alien, he was asked to stay where he was and then a large craft lifted off from the moor which disappeared. However, the analysis of the picture was not conclusive with suggestions of it being an unknown creature to its blurry nature. Strangely, based on a regressive hypnotherapy session, the retired officer Philip Spencer gave another account wherein he stated that while seeing the creature, he was instantly paralysed. He was then lifted into the craft and some other aliens examined him. He was then shown around the craft including two films, one of which was the impending disaster on this earth. The second film contents were not revealed since it seemed that the aliens did not want him to do so. There were the usual skeptics of the incident who claimed that it was a hoax and why no pictures were taken of the craft.

In 1994, Meng Zhaoguo reported an incident in the Phoenix mountains in China of having made contact with aliens after following a white shining object. He then reported that he was harassed by them and even forced to copulate in the spacecraft. He claimed that he was abducted from his home and shown Jupiter which was the purported home of the aliens. Investigations by Wuhan university revealed that probably the first encounter may have some element of truth but the rest were all concocted.

In 1997, Kirsan Illyumzhinov, a professional chess player, high ranking political figure and multi-millionaire in Russia claimed that he was abducted by aliens in yellow space suits from his apartment in Moscow. He further stated that he was taken to different planets and was told that chess was introduced to earth by them.

What do we make of these incidents? Without circumstantial evidence, it is difficult to validate any. However, many of the protagonists passed their psychological tests which revealed that they truly believed in what happened to them. Many were in a state of shock thereby alluding that something paranormal did occur but the question still remains as to whether these were actually contact with ETs. This may seem strange but then one cannot rule out some hysteria in the case of sighting of some unusual phenomena.

CHAPTER FIVE

UFO Sightings in Ancient and Medieval Periods

The Unidentified Flying Objects (UFO) have been reported ever since ancient history. Even some of the art of that period have captured some strange looking beings and crafts, not known to civilization at that point of time. Some of these include the rock paintings in Kanker, Chattisgarh, India where some characters even don clothing akin to spacesuits and the inhabitants have passed on the story of the small sized people visiting the in some round craft and taking people away. Similar artwork of the Mesopotamian and Egyptian civilisations also have some strange looking beings and craft.

However, one must admit that no one has really been able to vindicate it in terms of concrete evidence. The paranormal events surrounding many religions including presence of mythological creatures have also been the subject of much debate. There is a school of thought that believed that many of these events could be of extra-terrestrial origin.

On top of this there is another theory that aliens may have built many ancient structures such as the Pyramids of Egypt, monolithic statues of Easter Island, fortress of Sacsayhuamán in the Peruvian Andes, Stonehenge in

England, Teotihuacán's Pyramid outside Mexico city and many others. However, one must admit that no one has really been able to vindicate it in terms of concrete evidence which remains sketchy.

The paranormal events surrounding many religions including presence of mythological creatures, aerial modes of transport and celestial wars purported to involve advanced weapons have also been the subject of much debate. There is a school of thought that believed that many of these events could be of extra-terrestrial origin.

The Greek philosopher Metrodorus of Chios in 460 BC had said that

"To consider the Earth as the only populated world in infinite space is as absurd as to assert that in an entire field sown with millet, only one grain will grow"

The first historic sighting comes in 1440 BC of fiery discs in Ancient Egypt but the source, the Tulli Papyrus was disputed. Incidents were also reported from 218BC to 74 BC wherein strange phenomena such as crafts passing through the sky and drops of silver were reported but no further information is available on these. Some of these include the reported sightings of phantom ships in ancient Rome during 214BC by Titus Livus Patavinus a famous historian. Another Roman historian Plutarch in 74 BC has chronicled a huge flame like body falling between two warring armies. In the 1st century AD, Titus Flavius and any eyewitnesses saw a battle royale in the sky and this has been christened as the chariots in the sky.

As we come to the medieval period, there were a number of unexplained sightings. In 1015, some objects were seen to emerge out of stars in Kyoto, Japan. Around a couple of centuries later in Japan, some stars were seen circling in 1235. In 1461, the legal advisor to the Duke of Burgundy sights a spinning object that spiraled upward and then disappeared. In 1513, the famous renaissance painter Michaelangelo is supposed to have seen three lights with different colours in the sky.

In the 16th century, there were two celestial sightings in Nuremburg and Basel of aerial objects being flung in the sky. Reports in Nuremburg suggested that these objects were red semi-circular arcs to dark balls of ferrous colour and were flung from one end to the others. In Gangwon province of Korea in 1609, there were multiple witnesses in three different locations witnessing UFO in the form of a halo or washbowl. In 1638, three people saw a bright object appear over a river in Massachusetts wherein they even lost time.

In 1672 and later in 1686, astronomer Giovanni Cassini is said to spotted what he believed was a moon of Venus. However, there was no such moon and there is still speculation on what that object was. During an eclipse in 1820 in Paris, astronomer Fracois Arago saw a series of military style maneouvres made by UFOs.

However, all this sightings were never really investigated given the limited resources and relative primitive instruments at that point of time.

CHAPTER SIX

UFO Sightings until the 1950s

While there were incidents in the 19th century, many of these had little credibility. However, the sightings of mystery airships in the US from 1896-97 have still not been resolved. Such moving lights and objects were also seen in 1909 in Otago, New Zealand.

During World War II, there were sightings by pilots of metallic spheres and colourful balls of light. In this war period, there were tales of some sightings in Missouri and Los Angeles. However, there did not exist any physical proof and the alleged photo taken of aliens in the Missouri incident was lost.

The ghost rocket incident of 1946 in the Scandinavian countries of Finland and Sweden was another unique phenomenon. Initially, the strange rockets moving through the skies were thought of a meteors or even Russian tests of German missiles captured in the War. However, in the absence of any rocket debris found and strangely, these crashes occurring into lakes, an investigation was launched by both the Swedish and US military. Incidentally, such rocket crashes were also reported in Belgium, Greece, Italy and Portugal. The reports were inconclusive but there were hints at these rockets being of a technology not known on earth.

The sighting of 9 objects by Kenneth Arnold, a pilot in June, 1947 remains one of the most enigmatic. The theory of flying saucers came from this incident as the objects were convex shape with one in the form of a crescent. These were spotted in Washington State and were flying over Mount Rainer and going towards Mount Adams at an incredible speed of 1900 km/hour, faster than any known aircraft at that time. He ruled out any reflections due to the sun as he had rolled down his window. His hypothesis at that point of time was that it could be a guided missile or an advanced aircraft being developed by the US army. There was corroboration by a few people including an officer of the forest service on ground of these objects at that point of time. However, it was 10 days later that a United Airlines flight saw similar objects over Idaho.

Balloons have been the subject of much speculation over the UFO incidents. The Roswell incident of 1947 is one such where the debris of a crash was finally attributed to one such balloon in the state of New Mexico. However, quite interestingly, the Roswell Army had issued an initial press release wherein it was stated that a flying disc had crashed. They retracted this later and brought in the balloon theory. This remains one of the most enigmatic UFO incidents since it is believed by many experts that both the craft and aliens were found by the military which then investigated them.

Green fireballs have been witnessed in the US from 1948s into the 1950s, largely in the south western state of New Mexico. These were witnessed even by aircraft and it was attributed to meteors with some even claiming that it may be some tests near secret installations. There were other incidents in early 1948 like the crash of a P51 Mustang piloted by Captain Mantell while pursuing a light but this was attributed to a high altitude Skyhook experimental balloon which was a top secret project. Two pilots Chiles and Whittled in 1948 saw what they believed was a UFO in the shape of a cigar with windows and it whizzed past them. However, the subsequent air force investigation attributed it to a meteor.

The first photograph purportedly of a UFO was taken in May 1950 at a farm by a couple in McMinnville in Oregon, US. The two photographs were examined ad nauseum by a number of agencies with the skeptics pointing to the wrong time of sighting given by the couple and the possibility of the disc being a model, most likely a rear view mirror of a Ford which was hung from an electric wire. However, these suspicions also remained inconclusive and the couple maintained their story till their death.

The first video of a UFO was taken just three months later on 15 August, 1950 when Nick Mariana, a manager of a minor league baseball team in Great Falls, Montana filmed two bright silvery objects rotating and moving at a great speed over the stadium. His secretary was also there with him and both of them immediately saw two fighter jets also flying over the stadium after this incident which has been vindicated. The whole incident became a subject of much debate and controversy after Mariana claimed that the Air Force had removed some 35 frames from the original film which showed the rotation of the objects. There were a number of investigations on this film but the reports did not clearly rebut its genuineness. The film was also extensively publicized through documentaries and remains a solid piece of evidence that the discs were either something unknown or part of a secret mission given the follow up by the two fighter jets.

In the same month and year of the Mariana video, professors of the Texas Technological College in the town of Lubbock saw some lights in a U shape followed by similar sightings by residents with a student Carl Hart capturing some photos. The investigations initially pointed to flight of plovers who downsides were reflecting the town lights though many of the faculty rebutted that stating that the lights were much faster than a flight of a plover. The Lubbock lights thus remain a mystery.

One of the strangest UFO sightings remains that in July, 1952 of lights and balls of fire picked up by radars at the Washington National Airport right in the capital of the US. There were a series of these picked up and jets scrambled to search them. Some of the jets did not sight anything while the others could not match the speed of these lights which zoomed past them. It really spooked the Truman administration with so many sightings right in the heart of the capital and leading to a lot of speculation and public interest. Even come commercial airlines pilots and crew saw these lights around the city. Not surprisingly, high level investigations were conducted even by the Air Force and all the sightings debunked by the phenomena of temperature inversion wherein a layer of warm air over a cooler one could render the radar detection technology at that point of time faulty. The Air Force report even went into the extent of questioning the competence of the controllers at the airport. In the same month, two pilots spotted eight lights over Chesapeake Bay in West Virginia which was also corroborated by others on the ground. However, this was also rubbished by the investigative agencies as lights on the ground distorted by haze or fireflies trapped between the glass or reflection of Venus. Another similar incident was when two colonels Mc Ginn and Barton were flying their B-25 over Carson Sink in western Nevada and saw three aircraft which they initially though were F-86 jets. However, they were surprised by the altitude and the speed of these aircraft with a design without tails or a pilot's canopy. With reports of no other objects in that vicinity, the only plausible explanation could be some secret tests being carried out by the US government on a state of the art aircraft. The brouhaha created by these incidents also made the authorities ask the investigative agencies to tone down the importance given to these UFO sightings so as not to create public panic.

In August, 1952, a bluish white light was picked up near Haneda airbase in Tokyo and a F-94 was sent to track it. However, the light sped out of the range of the radar of the aircraft as soon as the tracking started.

One of the biggest UFO sightings outside the US was in a football stadium in Florence, Italy when around ten thousand spectators witnessed an UFO. It was in October,

1954 and there was a stunned silence when a group of UFO's purportedly cigar shaped stopped over the stadium as the match was under way. No explanations have been given of this incident and it still remains a mystery.

Owls are purported to have been held responsible for two alien sightings in the US. In September 1952, at the Braxton country of West Virginia, a group of children saw a bright light in the sky moving towards a property. They then narrated this incident and took others to search for the source. On reaching the place, some of them are reported to have seen a tall creature with blood red face, a hood, claws and draped in a hood making some hissing sound. A National guardsmen among the got frightened and dropped his flashlight leading to other scampering away. The incident shook them up but the investigations pointed the needed of suspicion to a barn owl which may have been frightened by the trespassing. The county however, christened this as the Flatwoods monster and even holds an annual festival.

The second of these was in Kentucky in August 1955 and is known as the Kelly-Hopkinsville encounter after the two towns where it happened. It was a bizzare incident in which a group of adults and children claimed that they were attacked by some around fifteen short dark aliens and had to fire to hold them off. When the law enforcement authorities reached, they could only find bullet riddled houses, a testament to the alleged encounter. Investigations pointed the suspicion to eagle owls or great horned owls which are large and can defend their nests with some aggression.

In November, 1957 in Levelland, Texas, many eyewitnesses saw a large ball of fire when they were driving and this temporarily shut down their engines when the object passed over their vehicles. An air force investigation which was not

comprehensive concluded that these were ball lightings or St Elmo's fire that shut out the electrical systems. However, some UFO experts pointed out that while there were rains, no lightning was reported at that time for the phenomena.

CHAPTER SEVEN

UFO Sightings from 1960s onwards

In April, 1964, a UFO was supposedly sighted by a police officer Lonnie Zamora while on a automobile chase in Soccora, New Mexico. He is supposed to have seen a white balloon and two short human shaped creatures in white overalls. However, the object then lifted off and left. The only two theories which fit the bill as part of the investigations were the lunar landing testing being carried out by the White Sands Missile Range and a prank by students of the New Mexico Tech.

July, 1965 saw another incident outside the US in Valensole, France wherein a farmer reported seeing two beings who paralysed him. He stated that they looked at the plants, made some grunting sounds and then returned to the vehicle. The only clue was some landing markings on the field. A very similar rural sighting in France was in August 1967 in Cussac, Cantal Province wherein a young sibling pair saw four alien beings around 4 feet tall. They were then reported to have floated up to a round spaceship, around 15 feet in diameter. The investigations revealed dry grass and sulphur odour in the ground where the reported sighting was made.

In September, 1965, the presence of two police officers in a UFO sighting in Exeter, New Hampshire is a difficult one to rebut. While it all started with the sightings by a teenager Norman Muscarello who was hitchhiking back to his home and spotted a large disc shaped object circling with five flashing red lights over two houses. He tried to call one of the families in the house but no one opened out of fear. He ducked away from the object which was coming towards him and took a lift to the nearest police station. With one officer Toland accompanying him, they went to the woods where the object was last sighted and saw it rise up again. They called for a backup and a second policeman Hunt also came in and just about managed to see the object before it flew off. This sighting was not surprisingly, a subject of much investigation with theories floated of possible refueling by a KC-97 tanker which had similar flashing lights, temperature inversion or lights from the nearby airbase. However, none of these seemed convincing either since a couple of law enforcement officers had seen a large UFO at low altitudes, almost on the ground.

In June 1972, a South Africa farmer Bennie Smit resident of Fort Beaufort in the Eastern Cape noticed an unidentified object hovering over his farm. He fired some shots with pinpoint accuracy but it had no effect on the object. Some police officers who came in later also fired at the object which finally lifted off and crashed into some nearby woods. It was investigated by an army regiment but subsequently had no records of the incident.

In January, 1975 near a high rise residential apartment known as Stonehenge in North Bergen, New Jersey, there was a sighting of an object by George O'Barski. He claimed to have seen ten hooded creatures who dis-embarked from the aircraft, collected some soil in a bag and then returned to the craft which subsequently moved away. The next day, he investigated the spot and saw some dug holes. There were other witnesses at the Stonehenge who also saw the aircraft.

In the next month, i.e. October, 1976, around a dozen lights were seen over Seoul. However, given the tensions with North Korea, the army fired anti-aircraft guns which injured many people on the ground due to the fallen bullets. However, the lights were unaffected by use of this weaponry.

The same year, 1976 saw an unusual incident where a group of four campers in Maine, US saw a bright yellow spherical ball over a lake in the Allagash area while paddling. While sending a torch signal to the spherical ball, a blue ray of light from the bottom of this light suddenly came over them. All of them only remember having seen their vessel dock on the shore and then they pitched their camp. Incidentally, they had seen the light a day earlier when there were other campers too. However, a couple of years after this camping incident, all four of them started getting nightmares wherein they saw these three ant headed humanoids who were peering at them as if they were being medically examined. All of them were subjected independently to a hypnotic regression where strangely, they recounted the same incident drew the same structure of the aliens. They also passed the lie detector tests. Many of the examiners were surprised and some even alluded to a para-normal experience of possible abduction that the four may have been subject to.

In September, 1977 around the Scandinavian countries and Northern Russia specifically over the town of Petrozavodsk, an array of lights was witnessed in the early hours of the day. These glowing objects had multiple sightings even in the cities of Copenhagen and Helsinki. In Petrozavodsk, it was a large light ball with multiple rays shooting. The investigations by the Soviet Academy of Sciences was inconclusive. Some of the theories floated were that of the launch of the Soviet satellite Kosmos 955, solar flares causing a magnetic perturbation, experiment of sending radio waves to the ionosphere causing this light phenomena etc. However, given the geography of the sightings, the suspicion fell on some Soviet test or launch.

December, 1978 saw a series of airline sightings of bright lights over the Kaikoura range in South Island, New Zealand. Basically, they were reported as large objects with white flashing lights which went on until 2015. The investigations pointed to many theories like light from squid boats reflecting off the clouds, unburnt meteors, light from Venus or even those from vehicles.

Subsequently, in December, 1980, a series of sightings in Rendlesham forest in Suffolk, England began with lights descending into the forest. However, the proximity of this place to two bases used by the Americans gave rise to some suspicion of some military vehicles. The investigative reports ranged from downed Soviet spy satellite to fireballs to atmospheric phenomena. It was also later reported that this could have been the handiwork of the British services to test the US bases preparedness. Some also pointed to the entire incident being a hoax.

Elsewhere in Texas, at the fag end of 1980, two women and a small child, were driving in an isolated road amidst dense woods when they saw a light in the shape of a diamond object which was emitting heat in the form of flames. As one of them began to observe it, it kept bobbing up and down and the heat became unbearable on account of which she rushed into the car. They reported that suddenly 23 Chinook helicopters came around the object and they all left. Two other witnesses also reported seeing some Chinook type helicopters but not the diamond shaped object. The two women suffered medically with nausea, vomiting, general weakness, burning sensation around the eyes and sunburns. The investigative reports did not conclude anything with some theories of it being the mirage of the star Canopus or the medical condition predating the incident. The US military denied the Chinooks or the diamond object were theirs. While it was baffling, the entire episode with the helicopters coming at the end, if true, and not related to ET, can only point to some secret military tests which obviously the establishment didn't want to own up.

In January, 1981, in the Trans en Provence, a farmer Renato Nicolai saw an object with the shape of two inverted saucers on top of each other land in a nearby field. When he went to inspect, he saw that the bottom had two objects protruding and the vehicle lifted off leaving burn marks where it had landed. When he reported this incident, the investigations by the police and a UFO investigation agency GEPAN did not come to any firm conclusions. The farmer believed it to be some military secret object given its proximity to a base.

In November, 1986, a Japanese Airline (JAL) cargo flight while entering Alaskan airspace from the north saw two cylindrical objects with lights that were moving at incredible speeds around the aircraft. With no air movement in the vicinity, the three crew members were stunned by that fact that the object transgressed all laws of gravity and were literally toying with the relatively slow placed aircraft. After sometime when approaching the city of Fairbanks, they saw another large object which the captain described as a huge mother ship. Nevertheless, a couple of aircraft in the vicinity did not pick up this object. However, the entire incident got captured on the radar data that was used in the investigation. Subsequently, in January, 1987, an Alaska Airline flight observed a fast moving object on its radar while the next day a US airforce plane also saw a large disc shaped object that soon disappeared. The entire analysis of this incident could not be disregarded like the other incidents given the radar data though some pointed out that the sightings came largely from the captain while the other crew only saw lights. In one of the US Department high level briefings, it was also stated that the incident should be kept a secret.

A series of sighting of lights in Belgium from November 1989 to April 1990 led to the scrambling of F-16 jet fighters to intercept it. However, the pilots reported no objects and locked each other while trying to do the same to the UFOs. A concoction thrown into this incident was a fake photograph circulated which was later owned up. The most plausible theory of these sightings was an atmospheric scattering known as the Bragg scattering.

In 1996, there were two incidents in Brazil of aliens and UFOs. The first of these was the alleged sighting of an alien by three girls in Varginha, who they described as being humanoid with no hair and red eyes. They ran away and reported the incident which sent shockwaves in the town. The skeptics dismissed this as the girls sighting a disheveled person and assuming he was an alien. There were rumours of another creature and the role of the military in picking it up. Reports suggested that the military had some aliens under its control. All this was rubbished in the investigations and the military also denied the same.

In the second incident at Saragonha island, a pilot Westindorff saw a large object which was around fifty to sixty metres high and circled around it. It was twisting and moving towards the sea. After some time, he saw a smaller similar shaped object come out of this larger one and disappear at a fast pace.

In March, 1997, a large number of eyewitnesses viewed a V shaped object with five lights move over Phoenix as also in the States of Nevada and Arizona. In a second event, a number of lights were visible which appeared at regular intervals and then disappeared. A number of photographs and videos were taken of the second event. The investigations primarily indicated that the first incident may have been an aircraft and the second one would have been military flares. However, there are still a lot of skeptics who believe that these were UFOs. The incident has been categorized as the Phoenix lights.

In 2004, some Indian scientists while camping in Chandratal, a lake in Himachal Pradesh, India witnessed a ball of light which moved much faster than any known objects . Many others have also spotted lights in this region with some investigators attributing it to weather balloons or drones. However, many witnesses have refuted this.

A strange incident occurred on 7 November, 2006 at the busy O'Hare airport in Chicago when around twelve eyewitnesses of United Airlines saw a saucer shaped object hovering over the airport for some time before shooting off through the clouds wherein a gaping hole was made. This was not picked up by air traffic controllers and even a proper investigation was not conducted with the Federal Aviation Authorities stating that it was a weather phenomena. This added to the mystery as the incident was widely covered given that it occurred at one of the world's busiest airports.

In a flight from Southampton, England to Alderney, France on 23 April, 2007, Ray Bowyer noticed two large yellow colour aircraft which were seemingly stationary. While they gave off a bright light, there was a black circular band at one end of these crafts. Although ground radar also picked it up, the second pilot Anderson said that this could have been atmospheric phenomena. However, on the return flight, the captain did not see the object. Interestingly, in January, 1994, an Airbus on its flight from Nice to London had observed a similar object near Paris.

Large commercial airlines have also noticed UFOs. On 13 July, 2013, an Airbus A320 while flying over Berkshire, UK noticed an object come towards the cockpit. Strangely the pilot thought that a collision was imminent but nothing happened. Later in 21 February, 2021, an American Airlines pilot flying over New Mexico reported a long cylindrical shaped object, akin to a cruise missile pass over them. However, this was not picked by air traffic controllers.

It is clear that some of these incidents do not have any explainable reasons. However, there is no evidence to also attribute it to ETs. It is important for us to adopt a more receptive attitude to the witnesses that come forth since it is only through such disclosures that we can really attempt to find the truth.

CHAPTER EIGHT

Sightings and narratives by children

This chapter is about sightings by children including school students of UFO and aliens. These are interesting for the fact that while one cannot rule out mass hysteria at such a young impressionable age, it is possible that given the number of witnesses and similar stories, there could be some element of truth. Moreover, it would difficult for large number of children to co-ordinate and fake any sighting. However, the skeptics do not rule out pranks when considering such incidents.

In April, 1966, around three hundred students and staff of the Westall High School, Melbourne, Australia reported seeing some flying saucers hover and then land in a field before taking off. The science teacher Andrew Greenwood also witnessed the UFO which he claimed was a white silvery object with a rod sticking up. However, his version was that the object was surrounded by five airplanes who were trying to track it down. However, the object whizzed at high speeds and after this cat and mouse game it vanished. Some children reported seeing three saucers. Some of the witnesses stated that the Headmaster and some other people, presumably from the Australian armed forces, had warned them not to speak on the incident. In an interview, given fifty years later in 2016, one of the students ruled out it being a military vehicle since it could do maneuvers which none of the modern transport vehicles could. Nevertheless, given the suspicion of the military trying to suppress this news and the presence of five planes encircling it, one cannot completely rule out some advanced military testing carried out.

The infamous Haven triangle was christened for a region in Broad Haven in Wales when in February, 1977, a cigar shaped vehicle with one silver colour creature was reported by students and later after a fortnight by teachers of a school. Even in nearby, Little Haven, there were sighting in the next month of an upside shaped disc and two humanoids. These incidents were never vindicated since there was little evidence thus attributing it to some prank.

A group of children playing football witnessed an unusual sighting in September, 1989 in the town of Voronezh in the erstwhile Soviet Union. It began with an initial ball of fire and around ten to twelve children seeing it. They also described a three eyed alien with bronze coloured boots coming out of the light and using a lazer gun to make a boy disappear. The police station of the town is also reported to have seen an unidentified object in the sky at that time. The investigations did not reach any outcome with some stating that it could have been the imagination of the children running wild.

On September, 1994, students of the Ariel School, in Ruwa, Zimbabwe were taking a recess break while the teachers were inside. The story goes that more than one silver craft landed near the school and short aliens, again numbering from one to four came out towards the children. One of them, dressed in black came close to the children and communicated telepathically a message about environmental protection and how humanity and its technology was destroying the planet. It was rural school with little exposure of the children to UFOs though there were sightings in nearby South Africa a few years before this incident was extensively reported. There was a flawed methodology in the initial interviews which were not conducted one to one. Some of the children also reported not having seen anything. In later interviews given, many children stuck to their positions while the teachers also mentioned that it was unlikely that the children would lie.

The case of Boriska Kipriyanovich, a boy born in 1996 in Volzhsky, Russia is as bizzare as the stories of alien abductions. Boriska was indeed a genius as he was able to read, write and draw at the age of around two. He had an amazing writing, language learning skills backed by an astonishing memory. He was also fascinated by space and especially the planet Mars. However, what was jawdropping was his claim to have come from Mars and being part of a martian race that was on the verge of destruction some thousand years back due to a nuclear war. He claimed that there were still some of his people on the planet who after the nuclear war had gone underground. He further stated that many of his race called the "indigo people" were sent to earth to warn people and save it from an impending catastrophe. Another of his bizzare claims is that given the interaction of his ancestors with the Egyptian civilization, there is a secret behind the ear to the opening up of the Sphinx in the pyramids and this would reveal something extraordinary which would change life on earth. Staying with our planet, he also said that the magnetic field of earth would switch and this could lead to an adverse environmental impact. While Boriska remains untraceable in Russia today, possibly taken to some undisclosed site, his claims do seem to be far-fetched given that if Mars had some civilisation a thousand years back, modern instruments should have identified some remnants even if a nuclear catastrophe had occurred.

While there are a lot of skeptics of these sightings, the strong arguments that exist are children from different families can rarely be excepted to concoct a mass lie and the witnesses were a significantly large number to brush it away. Moreover, many of the children, such as those in the Westall and Ariel school cases gave interviews as adults and were convinced that they saw something which the authorities didn't want them to divulge.

CHAPTER NINE

Area 51 and other secret bases

Area 51 is the nickname for the highly secret US Air Force facility in the State of Nevada and is administered by the Edwards Air Force Base. It is also known as Homey Airport or Groom Lake. There are no details available on this zone but the general understanding is that it is the training center for aircraft and weapons. Though not formally declared as a secret base, there is no information available on the research and incidents there.

One of the first projects in this facility was the testing of the U2 reconnaissance aircraft. The designer of U2, Kelly Johnson said that the dry bed of Groom lake was the ideal spot to land an aircraft for testing.

The next major project of Area 51 was the development of the Lockheed A-12 or the OXCART which was also a reconnaissance aircraft. The infrastructure was improved upon with a view to test such an aircraft. All this was completed in the period 1962-63.

Another project was of the Lockheed D21 drone testing. This was an ambitious project with the aim of reconnaissance on the military bases of cold war adversaries like the Soviet Union and China. However, in its standalone version, the drone ran into some rough weather due to an accident. Finally, it was dovetailed with the B52 bomber project. However, by that time satellite reconnaissance had become so advanced that the project had to be shelved.

Yet another project was the testing of US F-4 aircraft against the Soviet MIG 17 and MIG 21 aircraft which had been captured. The drill dogfights over Groom Lake gave the US an understanding of the capabilities of its adversary's aircraft.

Finally, the testing of stealth fighters like the Lockheed's F-117 Nighthawk was carried out in this zone. Moreover, the land around the area has been expanded including some highlands that were overlooking the base and could have posed a security concern for the secret testing that is carried out. There were some controversies surrounding this area on account of an environmental litigation and Skylab astronauts photographing this area from space.

On the face of it, Area 51 looks like any other secret base in a remote desert for testing. However, the secretiveness of the entire area in a country which has a vibrant democracy has sparked a lot of speculation. CIAs statement that this is the most secretive place on earth adds fuel to the fire.

In relation to ET, some of the theories range from examination of alien space craft or objects, meetings with ETs and research on time travel. During the 1950s and 1960s, there were a number of sighting of unidentified flying objects (UFOs). However, many of these were attributable to the testing of U2 and A12, both of which had superior speed, aerodynamic shape and larger bases reflecting sunlight. However, since there was no feedback from Area 51 operators on all the UFO sightings, one would never know the real story.

Some former workers in Area 51 have indicated that they were working on testing of alien aircraft. Some of them have also said that they worked with ET beings on telepathy and cloning of alien viruses. Robert Lazar, who purportedly worked in S4 zone of Area 51 gave an interview where he stated that he saw anti-gravity propulsion technology and a number of saucers. The authorities have however refuted all these allegations and have even disowned him as having worked there. However, this has never been verified independently and remains a subject of speculation. Nevertheless, all this only adds intrigue to Area 51 on whether it is limited to testing of advanced terrestrial equipment or does it also involve UFOs and ETs.

Kapustin Yar air base in the Soviet Union was yet another area which has been kept as a military secret. It was built during the end of World War II with a view to develop advanced technologies in military warfare. It is believed that the Soviet Union carried out a number of tests in this zone. Located close to the city of Volgograd, this base is also said to have a UFO test facility where the western intelligence alleges is a laboratory which tests UFO technologies and bodies of aliens. However, as usual, shrouded by complete secrecy, these are all speculation. Aerial photographs by the CIA have shown this base to have an intricate network of tunnels where these secret test facilities are supposed to be lodged.

Dulce base in New Mexico, US is suspected by some to be another secret underground base where alien technology may be tested. These allegations were made initially in 1979 by Paul Bennewitz who is a businessman based out of the area. Further allegations were made by George Andrews in his book "*Extra Terrestrials among us*" in 1986 and then by John Lear in 1987. Philip Schneider claimed that he was involved in the building of underground bases in the US including Dulce. The claim has been strengthened by many residents of the area reporting UFO sightings including strange moving lights.

While there are other bases in the world which are suspected by UFOlogists to harbor alien technology, the skeptics are not convinced. Moreover, many of these bases may be working on new technologies and hence have their access barred. The governments of US and Russia would not mind this enigma surrounding these bases since it keeps their potential adversaries in a state of uncertainty on military technologies.

CHAPTER TEN

Bermuda Triangle

Some may wonder if the Bermuda triangle and its enigma is apt for being discussed here when we are considering UFOs. However, as in all other conspiracy theories regarding aliens, this has also not be left out. The mystery began ever since the time of Christopher Columbus who saw a bright light crashing into the ocean when traversing the area in search of new lands in 1492. The purported paranormal activities in this watery area stretching in a triangle between Florida, Puerto Rico and Bermuda have been the subject of much speculation.

The believers of this school have attributed many reasons for these events. Some say that it is on account of the lost continent of Atlantis. Others have taken on Einstein's theory of relativity to state that this may be an area of a time space warp which is sucking in all these ships and planes. There is yet another explanation of some unknown and unexplained forces causing these events. There is also an attribution of compasses malfunctioning due to magnetic fields in the area. It is however, well known that this is one of the twelve areas of the world known as the vile vortices where the magnetic pull of the earth is high. This can cause interferences in the instrumentation, thereby sending vessels off route.

Of course people have also attributed natural causes to what is happening in the triangle. These include the area being prone to hurricanes. The Gulf stream has also been attributed to be causing some ocean going vessels to be swept away. Some have floated the theory that the ocean may have methane hydrates which have the property of making water frothy and not provide the buoyancy to ships for them to float.

Another conspicuous event was the March 1918 disappearance of the large US navy ship USS Cyclops carrying manganese ore with more than 300 of its crew which set off from Rio in Brazil. On its way to the eastern coast of US, it had to pass through the triangle. What is astonishing is that there was no distress call and no response to calls of other ships in the vicinity. The size of the ship and the number of people involved first led to a suspicion that the Germans may have sunk it in the war. However, the latter denied any knowledge even after World War 1 was over and the ship was never found. While the Bermuda triangle remained centrestage as the reason for the disappearance, some more plausible reasons were the overloading of the ship as well as a storm.

In December, 1945 when five Avenger torpedo bomber aircraft on a training flight in the region were lost. Strangely even the rescue mission in a PBM Mariner with thirteen men disappeared. The reason given by the Navy was a navigational error while in the case of the rescue mission, the Mariner was suspected to have exploded in line with a sighting by a tanker and earlier such incidents involving the plane.

There were however, a number of other incidents of loss of planes and even ships being lost or abandoned in the triangle. Some of these seaborne vessel were known as ghost ships since they were found floating without any crew and in some cases even with all the cargo intact.

A recent incident occurred in early 2021 when a boat with 20 people from Bahamas to Florida went missing. It was a 29-foot Mako Cuddy Cabin vessel. There were extensive searches carried out by the Coast Guard but this proved futile.

Despite the spate of accidents, in all honesty, the conspiracy theories have rarely focused on the alien or UFO angle despite many buying the argument that paranormal forces were at work. Nevertheless, unless the number of incidents comes down or the investigations provide a plausible reason for the event, there would remain speculation that the genesis could possibly be of an ET origin.

CHAPTER ELEVEN

Oumuamua and other strange objects

Interstellar space has always been a mystery to man despite all the technological advances. While advancements have been made in instruments with telescopes able to gaze at stars in a large swathe of our galaxy, man is still fairly primitive in terms of satellite technology. As discussed earlier, only five craft have managed to go beyond the solar system and are travelling in inter-stellar space. Moreover, they make take thousands of years to reach the nearest star.

It is in this context the visitor from interstellar space, "*Oumuamua*" generated so much of interest in 2017. A cigar shaped piece of rock, it came closest to the sun on 9 September, 2017 and was then discovered on 19 October, 2017 by Robert Weryk. Its dimensions were in a range of 100 to 1000 metres in length and 35 to 167 metres in both width and height. It was reddish in colour. In terms of motion, it is not rotating on its axis but tumbling along i.e. without any external force acting on it. Its composition is thought to be of dense metal rich rock and some speculate that the core could be of ice. Some scientists also stated that the core was made of nitrogen ice or hydrogen ice.

The genesis of the name comes from a Hawaiian term which translates as "*scout*". This was an object which seemingly came from inter-stellar space to scout for the solar system. For humanity, it was supposed to bring in information on the inter-stellar space beyond our solar system.

However, the first suspicion was that it may be a comet. However, this was ruled out since it did not have a coma which is basically a scattering when the comet is close to the sun. The second issue was whether it was an asteroid which are larger masses of rock orbiting the sun. The most famous of this are rocks in asteroid belt between Mars and Jupiter. However, the orbit of asteroids are generally within the solar system while this object came from interstellar space and was going back to this space.

The study carried out on *Oumuamua* seemed to suggest that as it neared the sun, there was an acceleration which could be attributed to solar radiation push. However, there was no outgassing i.e. ejection of gases due to the melting of the volatile particles with the heat of the sun. Outgassing is a phenomena shown by comets as they approach the star. However, as it moves out of the solar system, its speed had slowed down and it is likely to maintain the same unless it comes very close to any other star or a large planet.

Moreover, the trajectory of the object was more than elliptical and its speed was higher than that of the sun. These two pointers indicate that the genesis could be from inter-stellar space. One of the theories floated is that its origin could be from another star system, maybe even outside the Milky Way. Some also say that it could have come from a white dwarf, the last stage of a star. Yet another hypothesis is that it could be a fragmented part of a disrupted planet i.e. a planet that may have been struck by an object. Some also say that it could part of a disintegrated comet.

The final hypothesis was on the extra-terrestrial origin of *Oumuamua*. There were suggestions that the object may be a solar sail i.e. it uses the solar energy reflecting on mirrors for propulsion. It would also explain the sudden acceleration when it was near the sun. However, radio telescopes which were pointed towards it did not reveal any signals that could be attributed to its non-natural origin. Therefore, the ET origin of this object has little support from the scientific community.

When the "*New Horizons*" probe entered the Kuiper belt, just on the outer periphery of the solar system, it encountered a new object in 2016. A large rotating mass of rock by the name of "*Arawn*" was sought to be studied by the probe. However, when the probe neared the object, its sensors were cut off and no analysis could be undertaken. This was strange and led to suspicion on what this rock mass was? A rotating body was suspected by many to creating an artificial gravity inside which then pointed to its possibility of being a spaceship. However, given its speed of rotation which "*New Horizons*" picked up before shutting off, the theory is that if there were ETs inside, they would not be in humanoid form given the actual gravity being created. Surprisingly, as the probe moved away from *Arawn*, the communication systems were restored. This was indeed shocking since not the source of the communication failure could be attributed to this strange object, though there is no certainty, if this was due to a natural or external cause. Some of the ET protagonists were of the view that the Kuiper belt may be the ideal place for an alien civilization to observe earth, well hidden from telescopes and without being suspected. The belt has a number of objects which are believed to be icy objects, comets, dwarf planets and dust.

The objects such as the Oumuamua and Arawn will still carry the element of mystery until man improves his technological prowess and is able to have instrumentation that can given a conclusive verdict on it. Many more unknown objects may also be seen in the near future, especially with the operationalisation of the James Webb telescope. However, we must be cautious in rushing to any conclusions on its ET origin until some scientific proof or evidence exists.

CHAPTER TWELVE

Projects to monitor our universe

Before beginning this chapter which lists out the major projects that seek to scan the universe, we need to look briefly on the limitations of travel and observation of alien life. Firstly, knowledge of science as it exists today has limited the maximum speed of travel to that of light, i.e around 300,000 kilometres per second. This was propounded in Einstein's theory of relativity which also built in the concept of space time. However, what is astounding is that even with that speed, we would just be able to cover the nearest star in 4 years while the expanse of the universe would take around 14 billion years. It is a different proposition that the universe itself is expanding.

To put in perspective, the relatively snail paced space vehicles developed by mankind till date were the Apollo 10 with a speed of around 1/27000th of light while even the Parker solar probe would reach a speed assisted by sun's gravity of around 1/1600th that of light. Cracking the code for nearing or ideally speaking exceeding the speed of light is the greatest barrier to space exploration. Einstein in his study had postulated the concepts of worm holes, which could achieve the speeds of faster than light. These holes, if they exist are yet to be discovered by man.

However, in terms of communication with possible alien life, the situation is slightly rosier. The radio waves though discovered by Heinrich Hertz were put to practical used by Guiglimo Marconi in the form of radio transmitters and receivers. These waves can travel at the speed of light and hence could be a faster means of communication than space travel itself.

The Galileo Project, headed by Avi Loeb, a professor in Harvard University seeks to investigate extraterrestrial technological civilisations (ETCs). The aim of this project is to design new algorithms using data from astronomical surveys and telescope observations to identify potential interstellar travellers, alien-built satellites and unidentified aerial phenomena (UAP). It was announced through a press conference on 26 July, 2021

The personnel associated with this project would be multi-functional seeking to track inter-stellar movement though the use of telescopes stationed around the world. It would also focus on locating ETC satellites and analyzing unidentified aerial phenomena (UAP). The aim is to obtain high resolution and multi-detector images and analyzing them with computer models to uncover their source and nature. The project is to be based on validated transparent and systematic scientific research.

The project is a collaborative effort and seeks to pool in the efforts of the other sky gazers. One such source is Planet Labs which has a series of miniature satellites that scan the earth. The data from this source would be a valuable inputs for the Galileo Project, specifically to track and track UFO sightings.

The Project collaborators have also mentioned the need to look at the celestial body "*Oumuamua*" which was a cigar shaped object, which neither had characteristics of a comet or an asteroid. It came into our solar system in 2017 and stayed for a brief period of 2 months before moving out.

The James Webb telescope is the replacement for the Hubble Telescope and is a joint effort of NASA, the European Space Agency and Canadian Space Agency. It has been in January, 2022 positioned to start operations. It would use infra-red light to track objects in the universe. It is believed that this telescope with its state of the art equipment would be able to look at celestial objects in the universe which no other instrument has been able to. It has a large primary mirror which is around 6.5 meters and broken up into segments. Its instruments can detect even the faintest of signals. It would study the conditions from the formation of the universe to that of the solar system. Whether the telescope would be able to detect alien life is a big question but it would surely unravel many mysteries of the universe in its lifetime and provide man with a completely new perspective.

CHAPTER THIRTEEN

Possible life in the solar system

Among the moons, Jupiter's moon Europa, discovered by Galileo has been the subject of much study. It is made up of silicate rock and has an icy shell. It is believed that an ocean exists below it. There is speculation that the moon may have more water and oxygen than earth. Even the atmosphere including the exosphere of the moon is stated to be rich in oxygen. The Hubble telescope detected some water vapour plumes from the surface which indicates that there may be geysers from which the sub-terranean ocean may be spewing up this vapour. The ocean is supposed to be maintained on account of geological activity, tidal heating and irradiation. There is still no specific mission to land a craft on Europa but NASA's mission known as the Europa Clipper is likely to be launched in 2024 which would include a flyby of Europa.

Saturn's moon Enceladus was discovered by William Herschel has revealed traces of water geysers, quite similar to Europa. The spaceprobe Cassini discovered these geysers in the south pole with analysis suggesting an ocean with a thickness of around ten kilometres below the pole. What was interesting in this data was the presence of methane in these plumes of water vapour. This is also a harbinger of possible life since it could be emitted by micro-organisms in the ocean. Moreover, this vapour is also believed to have created the E-ring of Saturn which is composed of icy particles. It is suspected that the tidal heating or geothermal activity is supposed to have caused this oceans. Molecular hydrogen has also been detected on this moon which could also point towards the possibility of life.

Yet another suspect for harbouring life is Titan, the largest moon of Saturn which was discovered by Christiaan Huygens in 1655. The probability of ET life is accentuated by the presence of an atmosphere denser than the earth and an icy surface with possibility of a sub-terranean ocean. Moreover, it has both wind and rains which has led to the formation of lakes, rivers, deltas and even dunes. These rivers and lakes are largely made of hydrocarbons such as methane and ethane. All this came to light when the Cassini spacecraft made a flyby of the moon in 2004. However, there are many aspects that also deter this sense of optimism such as sub-zero temperatures on the surface, miniscule amount of sunlight penetrating the nearly 200 km atmosphere and the near absence of oxygen. Nevertheless, with all this information, one cannot rule out the presence of microbial or aquatic life on this moon.

Callisto, the second largest moon of Jupiter is a highly cratered body and was discovered by Galileo. It is composed of rocks and ice with an ocean below the surface. It has a thin atmosphere, believed to be composed of carbon-dioxide and oxygen. Moreover, being in the outer periphery of Jupiter, it has a low level of radiation and is thought to the best place around the gas giant for any possible future habitation.

Ganymede is the largest moon of Jupiter as well as of the solar system. It has a thin oxygen atmosphere which includes ozone. The metallic core is what probably leads to its own magnetic field. Studies by probes have indicated that its surface is made of silicate rocks and it may have oceans with more water than that on earth.

Triton, the largest moon of Neptune was discovered only in 1846 by William Lassell. Only Voyager II in its flypast had studied the moon which is known to be geologically active with some craters. It has an ice crust composed mainly of frozen nitrogen and a metallic core. The atmosphere is also largely nitrogenous. Interestingly, it has a retrograde orbit i.e. its orbit around Neptune is opposite to how the latter rotates. This is a sign that the moon may not be a natural satellite of Neptune but a dwarf planet which may have been captured by Neptune's gravitational pull while possibly slowing down due to a collision with one of its other moons. A school of thought is that Triton may have come from the Kuiper belt which is a mass of objects on the outer periphery of the solar system. Some also state that it may have been a binary system of the Kuiper belt which broke away. It has many similarities in structure and size to Pluto which is also part of the Kuiper belt.

Dione, a moon of Saturn was discovered by the astronomer Cassini in 1684. It is geologically active dotted with a number of ice cliffs, craters, ridges, canyons and geyser vents. While the core is rocky, the crust is composed of water ice with some suggestions of a large ocean below this. The atmosphere is very thin and is composed of ionic oxygen.

Charon, discovered in 1978 by a US naval station, is the natural satellite of Pluto. Ironically, Pluto is no longer a planet and is considered part of the Kuiper belt. The north pole of Charon consists of tholins which are organic molecules that could be the harbinger of life. They are thought to have been produced from methane, nitrogen and other gases which emanated from Pluto. It has a rocky core and an icy mantle and crust. There are water crystals and ammonia on the surface in the form of what is known as water ice. It is considered a binary planet system along with Pluto.

Lo is the nearest moon of Jupiter, is geologically the most active of all bodies with more than four hundred active volcanoes. The surface is silicate rock coated with sulphur and sulphur dioxide which are spewed from the vents of the volcanoes. The gravitational pull of Jupiter and its other moons has also caused high mountains on its surface. Due to its proximity to the gas giant, the radiation levels are very high. Therefore, the possibility of life on this moon looks the least likely as compared to the other moons that are a subject of study in this chapter.

However, one aspect that has often been overlooked in the quest for ET life is whether gases other than oxygen can also sustain life. Moving out of our traditional view of the elements that can harbor life, experiments were conducted on microbes if these could grow in pure hydrogen or methane. It was found that they did albeit at a slower pace. This is ample proof that we could also widen the scope of our search for microbial forms to other planets including the gas giants like Jupiter and Saturn which have an abundance of these two gases. However, given the high radiation levels in these two planets, it is not going to be easy to undertake studies. Moreover, the probability of life surviving at these radiation levels also gets minimized.

CHAPTER FOURTEEN

Planets that could harbor life

Are we the only body in the solar system harbouring life? In terms of the usual suspects of the planetary system, it is Mars and possibly Venus. However, some of the moons of other planets are also good candidates for some form of life, as we know it. The two planets are not surprising given their proximity to earth and thus harbouring the lesser harsher conditions to sustain life than possibly Mercury or the gas giants in the outer periphery of the solar system.

Venus has a thick cloud cover which does not allow sighting and study of the planetary surface. Interestingly in 1962, Soviet scientists had sent a Morse coded message to Venus in the first human attempt at inter-planetary communication. Scientists in September, 2020 using the Maxwell telescope in Hawaii and Atacama Array in Chile detected some phosphine in the upper clouds of the planet. This was a remarkable discovery and though the gas is poisonous, it can still harbor some bacteria. However, further scientific studies would be needed on the Venusian clouds to vindicate it as also gauge the source of this gas. However, the atmosphere of Venus is largely composed of carbon-dioxide which cannot sustain life as we know on earth. Venus is also known to have liquid water on its surface at the time of its formation. The flyby of the NASA's Parker solar probe in 2021 managed to capture images of Venus's red hot glowing surface which is said to be around 860 degrees celsius at night. The Magellan Mission in 1990 also provided input on the surface of Venus. These are expected to throw light on the planet's surface geology as well as composition of minerals thereof. All these are essential information for analyzing the possibility of life on the planet.

As far as Mars is concerned, studies indicate that this planet also had liquid water at some stage with currently only some small liquid brines left. This evidence comes from dried up river beds and freshwater lakes, polar ice caps, volcanoes and minerals. The martian atmosphere is said to have traces of methane, the source which could either be bio-organic or geo-thermal. The planet has always been colder than earth. With the loss of its magnetic field, its atmosphere also became thinner and radiation levels increased thus lowering the possibility of habitation. The land rover of "*Curiosity*" sent by NASA to the planet had indicated that the soil has the necessary life sustaining elements like carbon, hydrogen, nitrogen, oxygen, phosphorus and sulphur. The studies have focused on the sub-surface of the planet for signs of water. The presence of perchlorates and very low atmospheric pressures are also dampeners on sustainability of life on the planet. The presence of gullies on the surface of Mars have not conclusively proven these were created by liquid water in recent times. A chemical analysis of the martian meteorites such as Nakhla, Shergotty and Yamato000593 has led to speculation of organisms, but nothing has been conclusively proven. The analysis of geysers on the planet that regularly spew cold fluids with dark basaltic sand or mud have not led with any credible certainty on the presence of biological life. Scientists have also been working on studying bacteria which can survive in the inhospitable environs of the planet. The experiments have led to a belief that it may be difficult for micro-organisms to survive in these tough conditions. However, man has been seriously considering on setting up habitation zones on this planet given the presence of water. The use of latest scientific advancements without the need for humans such as robotics, artificial intelligence, 3 D printers, block chain etc would facilitate setting up of colonies. However, the raw data does suggest that the planet despite having very similar astronomical parameters like earth is still highly uninhabitable in its present climatic conditions.

The space probe "*Messenger*" studied Mercury in 2011 during its orbiting of the planet nearest to the sun. One of the most difficult tasks of the mission was the trajectory required to enter Mercury's orbit given the gravitational pull of the sun. It had thus to have one flyby of earth, two flybys of Venus and three flybys of Mercury itself before it could slow down its speeds adequately to enter the latter's orbit for the necessary studies to be undertaken. Water ice and organic compounds were found on the north pole of the planet, both positive signs of harbouring life. Moreover, water was also found on the exosphere of the planet. All this and some studies conducted in March, 2020 point to the possible existence of life at some stage of the planet's history.

We then come to our own satellite, the moon. There have many conspiracy theories on possible ET habitation on the dark side of the moon. This has begun with the Apollo 11 mission where it is believed that Buzz Aldrin, the second man on the moon saw an UFO. However, its weak atmosphere and the absence of water prevents life as we know it. However, the future missions could come back with greater geological understanding of its surface and whether it ever harboured life and are their pockets where life is possible?

CHAPTER FIFTEEN

Radio Signal communication to and from distant systems

In 1899, Nicola Tesla was working in his laboratory in Colorado Springs when he recorded some radio signals, purportedly from outer space. His hypothesis was that this was from an alien civilization more advanced than ours. Given the technology at that point of time which could not identify pulsars, most scientists were of the opinion that he may have detected one. Pulsars are radio signals given off by a collapsed large star known as neutron stars. Some say that the signals may have actually come from earth.

Carl Sagan in his article on Quest for Extraterrestrial intelligence in 1978 stated that "*A single message from space will show that it is possible to live through technological adolescence... It is possible that the future of human civilization depends on the receipt of interstellar messages*"

One of the most enigmatic events related to radio signals occurred on 15 August, 1977 when a strange signal was picked up by the Big Ear radio observatory in Ohio while on a search for ET intelligence as part of the SETI (Search for Extra Terrestrial Intelligence) project. This signal was believed to have come from the direction of the constellation of Sagittarius. It has been euphemistically called the WOW signal since the discoverer Jerry Ehrman had written the word "*Wow*" around the aberration in the radio wave readings that was responsible for this signal. However, strangely, there was no repeat despite sending a number of and huge volumes of signals in that direction. It is precisely this fact that was used as part of the investigations to conclude that the source may have actually been from earth, including from probably some light source or reflection off space debris. Some have also stated that hydrogen clouds surrounding two comets in that direction could also have been the source of the signal. However, some other scientists have disputed this having come from comets. The signal was however not picked up by observatories with more sensitive instruments. Whatever it was, this signal generated a lot of interest and subsequent analysis and some scientists still believe that this could be extra-terrestrial.

In 2017, radio communication was sent to one of the exo-planets GJ273b, three times the size of earth and which is around 12.5 light years away from us. This was done by an alien hunting group called the Messaging Extraterrestrial Intelligence International (METI). The GJB273, also known as Lutyen's star is a red dwarf star with two planets around it. One of these planets is presumed to have conditions very similar to earth and the message could take more than twelve years to reach.

However, it is interesting to note that there are other exo-planets in our galaxy which have very similar conditions to our planet and could be potential candidates for life. Some of these, in the order of distance from earth include the Tau Citi e and f which are 11.9 light years away, Gliese 581g and 581 d which are 20 light years away, Gliese 667Cc which is 22 light years away, HD85512b which is 35 light years away, Gliese 163c which is 50 light years away, HD 40307g which is 52 light years away, Kepler186f which is around 480 light years away and Kepler 22b which is 600 light years away. However, communication with these exo-planets is yet to be made.

In December, 2020; scientists captured strange radio signals from the nearest star to our solar system, the Alpha Centauri. The Alpha Centauri system consists of three stars Proxima Centauri which is the closest star to the sun as well as Alpha Centauri A and Alpha Centauri B. This electromagnetic beam at 980 megahertz was speculated to indicate a source of life from probably a planet which is around 17% larger than earth and could be in a habitable zone from the star. There is a possibility that this planet could have water, one of the essential sources of life. However, this star is still some 4.2 light years away, something which would take even our fastest spaceships around 17,000 years to reach. The source of these signals has still thrown up many options ranging from a comet to a hydrogen cloud to even some human technology from some unknown source on the planet. However, while dwelling on the Alpha Centauri, one cannot forget the alleged 1956 abduction incident of Elizabeth Klarer in which she narrated having gone to the planet Meton which supposedly circles the Alpha Centauri. Nevertheless, to put in perspective, most UFOligists believe that it was concocted story and a hoax.

The reception of radio signals from distant galaxies is still work in progress. Humans still don't comprehend all these radio signals being transmitted. For example in January, 2022; some astronomers using the Muhcinson Widefield Array in Western Australia discovered a regular burst of radio signals every 18.2 minutes. This was coming from a system nearly 4000 lights years away. The fact that this was being transmitted at different frequencies could be an indication that it is natural. Some believe that this maybe from a white dwarf.

The famous American astrophysicist Neil de Grasse Tyson once stated:

FM signals and those of broadcast television...[travel] out to space at the speed of light. Any eavesdropping alien civilization will know all about our TV programs (probably a bad thing), will hear all our FM music (probably a good thing), and know nothing of the politics of AM talk-show hosts (probably a safe thing)

It is interesting to note that the first radio waves sent from earth have travelled only around two hundred light years which means that another 120,000 years would be taken for them to even traverse the entire milkyway. Hence, communication with radio waves would be reciprocated only if the aliens are very close to us or have very advanced technologies that can pick up these radio waves even from distant places.

CHAPTER SIXTEEN

Inferences

What do all the sightings, interviews and statements on UFOs lead us to? Some may say, not much as we are yet not clear on the existence of ETs. Nevertheless, many of the sightings and event have not been explained given the reports from investigators who believe that the subjects believed in what they saw and reported. However, it is important to understand from the perspective of the powers that be, who may or may not be hiding some information from humanity? It is also difficult to gauge as to who are these powers? Is it the US or are some other powerful countries also in the so called loop?

The science fiction author Arthur Clarke had mentioned that

"Two possibilities exist: Either we are alone in the Universe or we are not. Both are equally terrifying."

There could only be a few scenarios that we need to explore and understand. However, all this is pure speculation based on parameters of the ETs such as their intelligence level and whether they have indeed found us.

Firstly, let us assume that the ETs have been found and they are much more intelligent than us? In such a scenario, which most UFOlogists tend to believe in, wouldn't it be a case where that advanced species would be calling the shots rather than the powers that be on earth. They would surely not allow the earthlings, whether the US or Russia or China or others, to decide as to what to do and whether the humans should be told about their existence. They may have technologies which they would decide it if worthy of being shared with any country on earth. They can very well take over the military bases and virtually control, even the powerful countries. Even if they are not the predators that Stephen Hawking and others talked about, surely, they won't be coming to earth for some charity, and that too to help out some select country with their technology and then return to where they came from. Another offshoot of this theory is these advanced species are just watching us, not wanting to interact, for whatever reason it may be. However, in that case, for all purposes, even the powers that be are as clueless as the ordinary denizens of this planet.

The next scenario is that the ETs who came here are just about as or lesser advanced than us. The probability is low since in that case, it is unlikely that they would have found us. How would they have travelled all the way to earth when even for mankind based on our latest technologies, it would take us nearly 20,000 years to reach the nearest star, the Alpha Centauri system. But assuming that they did due to some technology which we are not privy to, then can they really be less advanced in other aspects. That one technology of theirs, which allows them to travel so fast would virtually place them on a pedestal of being more advanced. But assuming that the ETs are present in some human bases, maybe Area 51 or some other secret location, why would they stay on here? Is it just to interact with our race, understand us or collaborate with us? While all this is pure speculation, it doesn't seem logical that they would be stay on for long when they have the technology to go back from where they came from.

The third scenario, is that we have just communicated with some far off ETs but given the limitations of existing technology, haven't been able to re-establish contact. This is on account of time period of travel of radio waves being finite and the distances being too enormous. This would call for resumption of the communication channels to understand each other including the locations. However, this does not explain the series of alleged sightings and the entire UFO paranoia.

Yet another scenario is that an advanced species of ETs is responsible for all the sightings and alleged abductions. They may be possibly conducting experiments or just observing our advancement in the technology space. They may be creating an alien-human hybrid as many of the subjects have reported and our powers that be are aware of this but cannot do anything. This looks a scary scenario but then it doesn't make sense for them to be playing this cat and mouse game of observing us and our technologies for so many years. What would they get out of having these hybrids? Would it ensure the longevity of their race? Moreover, they could very well control us and achieve whatever objective they have set out for.

The last scenario is that we haven't established any contact with any ET and all the existing communication's received and sightings made are still to be vindicated or could be actually be some unexplained natural phenomena. Maybe even the unexplained sightings could be something beyond the understanding of our science. This would surely be a disappointing option for most UFO hunters but could actually be the most realistic one. The series of interviews of political leaders looks to be pointing towards this and the general comment made by some of them is that while we do not understand certain UFO phenomena but with the presence of billion of planets in a habitable zone, one can't rule out other life forms in the universe. This could be inferred a general statement which is more an outcome of frustration that no ET life has been conclusively found. The next question that comes to mind which has been dealt earlier in this book is the rationale for keeping this under wraps. Well the only logical reason could be that the UFO phenomena serves an important purpose of keeping your potential adversaries on a tenterhook, uncertain on the technologies that you might possess from ETs. It also keeps the enthusiasts glued onto the subject and more importantly ensures a continuous flow of funds for space research. So why burst this bubble and create a wave of disappointment? It surely serves no purpose.

CHAPTER SEVENTEEN

Conclusions

As the enigma of ET life persists, it is time to reflect on the plethora of events recorded and whether we will ever make contact with life outside our planet. Some say not during our lifetime unless we are lucky, but definitely during some stage of human advancement. Some of the UFOlists say that we may have already done so given the series of unexplained encounters.

What is clear from the various sightings and reporting are that there is scant evidence of ET life even though many of the sightings remain a mystery despite some theories like hallucinations, atmospheric phenomena, top secret military hardware, spy gadgets, weather balloons, satellites and even possible pranks. Some of the people who saw the UFOs and aliens seem to be convinced that they did see something paranormal and have come out with their theories despite the stigma of being branded.

A school of thought could be that these are all fake encounters and is humanity yet to make any contact? With the US administration releasing the radar images of those unexplained objects that were hovering over their state of the art warships and destroyers, one would find it hard to categorically say that these are not ETs. Nevertheless, Stephen Hawking in one of his quotes gave us a stark

message on all this phenomena

"If the government is covering up knowledge of aliens, they are doing a better job of it than they do at anything else."

However, it is a fact that with so many star systems in our universe, even the basic theory of probability should have ensured that a number of planetary bodies and their satellites would meet the criteria of conditions ripe for life, as we know it on earth. But even if they do exist, the question arises of the level of evolution of the civilisations therein. Unless they are more advanced than what man is today, it is unlikely that they would be able to contact us with the astronomical distances involved in any commute or even through advanced telescopes.

Another theory that is floated is that it is possible that there are advanced alien civilisations and they just do not want to contact us. Maybe they are observing us but do not see it as opportune to communicate with us, whatever be the reason. Apart from the fact that they may not want to disturb the current equilibrium by contacting us, they may be living a sustainable life's themselves to worry about galactic explorations.

Some postulate that aliens may be trying to contact us but probably because of their advanced technologies that cut down on time taken for communication, we may not have the wherewithal to receive these messages. Or it could well be that they may be at the same level of progression as or primitive than ours?

Another aspect is that we may not be able to comprehend the life forms apart from ours. While looking for tell-tale signs of life giving elements like hydrocarbons, water, atmosphere, oxygen; life could exist in other forms and in the absence of many of these elements. Thus ET life could be in a completely different dimension, shape, size, composition and intelligence. Maybe, we are too primitive to even understand it and may not be able to decipher their multi-dimensional presence among us.

Just like the analogy that we may not be able to comprehend other life forms, there is also a possibility that mankind is too primitive to understand the laws that govern universe. Our understanding of physics and forces may be limited and may not be applicable to other parts of the universe. There would be technologies that may completely alter our understanding of the subject and enable us to overcome the many barriers that limit our speeds for exploration of the universe like gravity, matter etc.

Some also say that the aliens may have come prior to our evolution on this planet. It is possible that they may have explored the planet and noted its position for possible future communication. Another school of thought also believes that they may have sown the seeds of our civilization and we all may be the subject of an experiment.

An interesting school of thought from some scientists like Carl Sagan and Stephen Hawking is that there may be powerful aliens who are predators and could destroy all other civilizations. Our quest to contact aliens may thus prove counterproductive and we could be attacked by these predators. The existence of these predators could also be a reason that other advanced civilisations may not be contacting others in the universe.

Some others believe that we may be located in one unexplored part of the universe which has not been contacted or colonized. The proponents believe that the rest of the universe may have been conquered by other civilisations and we may thus be just lucky due to our remote location in the universe.

It is a fact that man has been seriously exploring the universe with his limited technologies only for the past hundred years. This is too short a time period in the context of the age of planetary systems and that of the universe itself. It is possible that we are outside the periodicity of contact being made by other advanced civilisations.

So what should man do? He would definitely need to explore the universe within the limitations of his technology. Maybe, we may catch some signals of ET life somewhere outside our system or even the galaxy. Even the existing human satellites that are out of the solar system would take many years to reach even the nearest stars. This definitely dims the ray of optimism but then some advanced civilisations would well try to contact these satellites. Of course the final alignment of the James Webb telescope should open up a new vista for us to detect the furthest of objects even in the remote corners of our universe. It should provide us a completely different perspective with the images beamed back.

We also need to be more receptive to people reporting about UFOs and aliens. While it carries a risk of increase in hoax reporting, there is no other option, if we need to get to the roots of the unexplained encounters. Stigmatization is something we must avoid.

As far as the solar system goes, it won't be surprising that we may find some microbial life in some of the planets or moons especially if we are able to detect and explore the large oceans of water and even the atmosphere of some of the suspect bodies dealt with in this book. One can safely stick their necks out to say that this may not be far away given the number of probes being planned to these planets and moon and the relative advancements in our technology to analyse them.

As for UFOs and alien life beyond our solar system, the biggest challenge is speed of travel and communication. Given the vastness of the universe, which is also expanding, we would need to really upskill in terms of technology. However, even Einstein in his theory of relativity had pointed to the existence of worm holes that could take us from one

point of the universe to the other beyond the speed of light. Our present understanding of science is very limited and unless we can really overcome the known barriers to space travel chiefly cracking the speed of light, it may be a long time before we could think of long inter-stellar exploration. However, the solar sail technology can also be tinkered to get speeds slowly up to that of light. Even if that is mastered, man could consider atleast moving out of the solar system in the first phase.

Nevertheless, even with his primitive technology, man would surely continue his scouting for ET life. The communication with radio signals remains currently our best shot until there is a technological breakthrough of inter-stellar travel. This is surely driven by the optimism that there is ET life somewhere in that vastness of spacetime. After all "*If it is just us, seems like an awful waste of space.*" said Carl Sagan.

Printed by Libri Plureos GmbH in Hamburg,
Germany